BAOFENG FOR BEGINNERS: Your Easy Radio Guide

Russell Clarence

1

Table of Contents

CHAPTER ONE

INTRODUCTION

You're examining up for your Lord Class FCC award with Ham Radio Prep's web-based construction or you truly got endorsed and eventually you truly need to want to convey progressively! What's the most reasonable and quickest method for managing getting on the ham social events? You're checking the headways from novice radio vendors and finding that principal VHF and UHF amateur handheld and adaptable radios cost some spot in the extent of $100 to as much as various dollars. Is there a strategy for getting on the ham hands on a tight spending plan?

The present overall business put awards you to get on the air for unobtrusively, the responsibilities accessible can show vacillating.

CHINESE AND IMPORT RADIOS OFFER A BENEFIT FOR NEW HAMS

The basic producers of ham radio handsets have been around a shockingly prolonged stretch of time: Yaesu, Icom, Kenwood and Alinco. Regardless, the general business local area in this way by and by has players like Baofeng, Wouxun, Anytone, Retevis and altogether more names. While their overall market solidifies two-way radio clients out of control, these world makers and backers likewise comprehend that their things have tracked down their bearing into the hearts of hams in the US, yet different region of the world, too. While you can pay $100 to $200 for a ham handheld radio made by the ham radio colossal four, you besides can get on the air effectively and humbly with a key import handheld radio for as low as $12! Did we say $12? Without a doubt!

UTILIZE A PC TO PROGRAM YOUR RADIO

The Baofeng BF-888S, which looks like the Retevis H777, sells for some spot in the extent of $10 to as much as $25 and offers 16 channels on UHF, which for hams would be the 440-450 MHz band where most repeater and simplex correspondences happen in the by and large 420-450 MHz USA ham band. You'll require a programming join (sells for $5 to $15) and a PC with the radio's free programming open for download online to program the handset.

A drive ahead off the single-band UHF handheld is a twofold band radio, for example, the Baofeng UV-5R handheld (and its different subordinates), the most famous of all Chinese and import radios. This radio sells online for some spot in the extent of $20 to $35 and offers full thought of the ham 2-meter and 440-MHz social affairs. The limitation is that these radios similarly will program on business and public flourishing frequencies, expecting you have such a need for twofold responsibility. While radios, for example, UV-5R or sister UV-82 can be revamped by hand instead of utilizing a PC, you will find it considerably clearer to utilize the programming get together with a PC to get different frequencies changed for neighborhood association. You'll need to program all your nearby VHF and UHF repeaters, as well as simplex frequencies like 146.520 and 446.000 MHz.

ARE BAOFENG RADIOS ACCEPTABLE FOR HAM USE

Ask concerning whether a Baofeng radio is Acceptable for use on ham radio and you will hear two reactions:

1) No, taking into account how they are subtle waste, and

2) Sure, I have one myself.

You'll see that hams are either possibly in favor of import radios, subject to their examination of them. Take the necessary steps not to let that influence you. Go with your spending plan. In the event that you can manage the cost of one of the monstrous four ham rigs, put all that in danger. Expecting you need to set aside some cash, go with an import radio and perceive how it abilities for you. You can ceaselessly overhaul later to ham stuff or even business two-way gear (which is certifiable on ham social occasions). Nothing horrible can be said about consuming $25 to ponder making the jump. You'd be shocked to find many hams own an import radio either for their key radio, a help radio or notwithstanding, for their bugout unit for crisis reaction.

"Your situation will be unprecedented" and the power result of import handhelds also may differentiate. The Baofeng BF-888S handheld may be progressed as having 5 watts of force, yet you'll more conceivable see a power result of some spot in the extent of 1.5 to 3 watts. The UV-5R conventionally is progressed as 5 watts yield, yet you could consider it to be additional like 4 watts. Be careful of varieties that confirmation power eventual outcome of 10-15 watts since you'll essentially 100 percent see only 8 watts.

Ham radio shouldn't mess around with to be an extravagant hypothesis for new FCC Capable class licensees! Spend only a tad and get on the air rapidly. Assuming it works for you, get a flexible radio wire or a pervasive handheld radio wire, an extra battery, a 12-volt vehicle connector or a handheld recipient all out for not the actual expense of a genuine ham handheld expecting you somehow wound up buying that enormous number of embellishments.

Orders, Keys and Fastens

PTT (PUSH-TO-TALK): Press and hold to send, discharge it to get.

SIDE KEY1/[CALL]: Press to begin the FM radio. Press it again to deactivate. Press and hold to begin the watchfulness button. Hold again to deactivate.

SIDE KEY2/ [MONI]: Press to turn on the electric light. Hold the strategy for screening the sign (quickly handicap squash).

[VFO/MR] BUTTON: Press to switch between pre-changed channel modes, and rehash mode.

[A/B] BUTTON: Press to switch the recurrent show. This will figure out which of the two showed frequencies you're sending and getting on.

[BAND] BUTTON: Press to switch the band you're managing, which is wilt VHF (136 mHz) or UHF (470 mHz). While the

FM radio is instigated, press to switch the FM radio social affairs (65-75MHZ or 76-108 MHZ).

[SCAN] KEY: Hold the button for two seconds to begin taking a gander at for dynamic channels (channels that are granting). The radio will in this way stop at a rehash in the event that it sees action. While the FM radio is dynamic, hold to look for radio broadcasts.

["KEY"] KEY: Press while in Channel mode, press to switch among High and Low bestows power. Press and hold for two seconds to lock and open the keypad. This is valuable for when the radio is on and you need to get correspondences, yet you truly need to store the radio without button-pounding any settings.

[MENU] KEY: Press to enter the focal menu and to pick and affirm settings (goes about as "ENTER" key).

Bolt KEYS: Press and hold the UP or DOWN bolt keys to dial the rehash or adjusted channels up or down while not in the menu. Utilize the bolts to examine the menu, moreover.

[EXIT] KEY: Press to drop a limit or leave a menu or screen.

CHAPTER TWO

LITTLE BY LITTLE BEARINGS TO ENTER A CHIEF CRISIS REHASH

You can utilize the UV-5R quickly. It just stops momentarily to turn it on and set up a significant rehash or channel to send and get. Since this is genuine your most vital time utilizing the radio, might we at any point go over major game-plan first.

RESET/"ZERO OUT THE RADIO

To strategy the radio, guarantee the battery pack is snapped to the rear of the handset. String the receiving wire onto the receiving wire post and fix. Turn the radio on by turning the volume handle clockwise. It'll click, the radio will blast two times, and in this way a voice will state "Rehash Mode" or "Channel Mode".

TIP: You ought to zero out (reset) the radio to its default settings to guarantee there are no planned settings which could frustrate crisis correspondences:

1. Press MENU.

2. Use the all over bolts on the keypad to examine to menu choice 40.

3. Press MENU again to pick "ALL".

4. Press MENU a third opportunity to pick "SOURCE?"

5. Press MENU a fourth opportunity to reset the radio.

SELECT YOUR INCLINED IN THE DIRECTION OF LANGUAGE

The radio will reset and default to a Chinese voice. To pick you're inclined in the direction of language:

1. Press MENU.

2. Navigate to menu choice 14.

3. Press MENU again to pick the language affirmation.

4. Use the bolts to find "ENG" for English (or you're leaned toward language).

5. Press MENU again to declare the language confirmation.

6.   Exit.

UTILIZE THE UV-5R AS A FM RADIO

The UV-5R's most essential limit is behaving like a FM radio for your 1 station. This is valuable during calamities, when crisis broadcasts and data are put out through adjoining radio broadcasts. To draw in FM mode, essentially press the orange "CALL" button on the size of the radio. Check each station open by endlessly smashing the */"Clear" key.

Enter, save, and utilize a crisis rehash

You can enter a rehash and begin getting and sending by basically framing the sensible numbers on the keypad. For instance, making in 162.400 will enter you into the NOAA weather conditions broadcast. Framing in 151.940 will enter you into the most comprehensively seen public crisis channel. We truly need to program some crisis channels so we don't need to review each of the digits to each recurrent we could utilize. To save a rehash and make another channel:

1. Press VFO/MR to place the radio into Rehash (VFO) Mode.

2. Press the A/B button to choose the top recurrence. Note the bolt to one side of the recurrence on the showcase, demonstrating your choice. All

programming should be finished utilizing the top recurrence.

3. Turn off TDR/Double Backup (it ought to be off however affirming it is).

Press MENU.

Press 7.

Press MENU to choose the menu choice.

Use the all over bolts to choose "OFF".

Press Menu to affirm.

Exit.

4. Type in the recurrence you need to save utilizing the keypad.

5. Press MENU.

6. Navigate to choice 27.

7. Press MENU again to enter the channel determination.

8. Select the ideal channel (000 to 127) by squeezing the all over bolts. We suggest beginning at channel 1, then, at that point, 2, etc. In the event that a channel number has "CH-" before it, that channel as of now has a recurrence saved.

9. Press MENU to save the recurrence to the chose channel.

10. Exit.

You can now choose the saved recurrence by squeezing VFO/MR to choose Channel Mode, and afterward squeezing the all over bolts. The radio will burn through every one of the saved frequencies' channels.

While in Channel Mode, the presentation will show two of your saved frequencies, and the channel every recurrence is saved money on.

ERASE A SAVED RECURRENCE CHANNEL

Erasing a recurrence or channel is much simpler:

1. Press MENU.

2. Navigate to choice 28.

3. Press MENU to enter channel determination.

4. Select the channel/recurrence you wish to erase.

5. Press MENU again to erase it.

6. Exit.

CHAPTER THREE

LOOK FOR DYNAMIC FREQUENCIES AND TRANSMISSIONS

You might wind up in a fiasco circumstance with no known frequencies or channels. Assuming this occurs, you can in any case utilize the UV-5R to get crisis correspondences by utilizing it to check the wireless transmissions:

1. Press VFO/MR and guarantee the radio is in Recurrence Mode.

2. Press and hold the */"Output" Key.

3. The radio will look over frequencies quickly, halting when it hears a transmission.

4. To set the quantity of frequencies the radio leaps with each output, press MENU.

5. Navigate to choice 1: "STEP".

6. Press MENU to enter the step choice.

7. Use the all over bolts to increment or decline how much the step.

8. The least step (2.5K) is the slowest and most exhaustive pursuit. 50K is the quickest and least intensive recurrence search.

CHANGE THE RADIO'S WORKING BAND VHF OR UHF

The Baofeng works in two groups: Extremely High and Ultra High Recurrence. Only one of two groups can be checked and utilized at a time. To switch between groups:

1. Press MENU.

2. Navigate to choice 33: BAND.

3. Press MENU again to enter the band choice.

4. Use the all over bolts to choose VHF or UHF.

5. Press MENU again to affirm.

6. Exit.

TIP: Numerous crisis radio frequencies, police, EMS, government organizations, and salvage tasks utilize the VHF band.

CTCSS AND DCS PRIVATE LINE OR PL INTERCHANGES

Now and again, one radio recurrence will be utilized for communicating and getting by different administrators. This is particularly possible in a fiasco situation.

Numerous people on call, war rooms, and salvage tasks will adhere to one recurrence to guarantee stable correspondences.

However, to isolate every one of the administrators on a solitary recurrence from one another (and to abstain from sharing transmissions and swarming up the wireless transmissions), two potential frameworks of apparent frequencies are utilized.

These apparent recurrence frameworks are called CTCSS (Ceaseless Tone Coded Suppress Framework) and DCS (Computerized Code Crush). The two frameworks work to a great extent the equivalent, with the exception of DCS is computerized.

• CTCSS has 50 widespread tones estimated in hertz (67.0 Hz)

• DCS has 105 all inclusive tones estimated alphanumerically (D023N)

To have a superior comprehension of this, consider a remote telephone and the cell network it works on. Consider the actual recurrence the organization, and the CTCSS or DCS tone as the telephone number. You should be on the organization to send or get, yet you likewise need to "dial" the right number to impart (the CTCSS or DCS). There are all inclusive, prearranged tones incorporated into the UV-5R for the two frameworks. You should realize which tone the recurrence being referred to is utilizing, to communicate.

INSTRUCTIONS TO PROGRAM CTCSS AND DCS TO A RECURRENCE CHANNEL

To program a CTCSS or DCS tone into a recurrence that requires it (and save it to a channel):

1. Press VFO/MR and put the radio in Channel Mode.

2. Ensure you're on Channel A by squeezing A/B.

3. Type in the recurrence you need to save.

4. Press MENU.

5. Navigate to choice 10 and 12 to set a sending and getting DCS tone.

6. Navigate to choice 11 and 13 select a sending and getting CTCS tone.

7. Press MENU again to choose the proper choice.

8. Use the all over bolt keys to choose the fitting DCS or CTCS communicating and getting tones. In choices 10 and 12, or 11 and 13.

9. Press MENU to affirm your choice for each.

10. Navigate to choice 27 to store the recurrence and the communicating and getting DCS or CTCS tones to a channel.

11. Exit.

Presently, the recurrence you just saved ought to be on the presentation with all things considered "DCS" or "CT" to one side.

EFFECTIVE METHOD TO PROGRAM A REPEATER RECURRENCE

HAM radios can travel specific distances straightforwardly, normally a couple of miles. Repeaters carry on like "waypoints" for a transmission, bouncing it from one repeater to next, expanding your correspondence range. This is staggeringly helpful in a debacle circumstance since it permits you to arrive at a lot more prominent distances. A few repeaters permit transmissions to arrive at huge number of miles.

To program a repeater and communicate on its recurrence, you'll have to know some data about the actual repeater:

• Repeater recurrence

• Shift (+ or -)

• Balance

• R-CTCS or R-DCS (once in a while)

• T-CTCS or T-DCS

A few repeaters utilize different apparent recurrence frameworks that are not CTCS or DCS, or none by any means. Those different frameworks can't be customized into the UV-5R. They are not canvassed in this aide.

To program a repeater:

1. Press VFO/MR and put the radio in Channel Mode.

2. Ensure you're on Channel A by squeezing A/B.

3. Type in the recurrence for the repeater you need to save.

4. Press MENU.

5. Navigate to choice 10 or 11 to include the R-DCS or R-CTCS (if appropriate).

6. Navigate to choice 12 to 13 to include the T-CTCS or T-DCS.

7. Navigate to choice 25: SFT-D.

8. Set the positive or negative shift for the CTCS/DCS (gave).

9. Navigate to choice 26: Offset.

10. Set the proper offset (gave in view of band).

11. Navigate to choice 27 and save your repeater to a channel.

12. Exit.

Presently, in Channel mode, you can choose the repeater channel and communicate on the repeater being referred to.

CHAPTER FOUR

HELPFUL CRISIS RADIO FREQUENCIES

Presently you have an essential comprehension and can program and utilize the UV-5R. Here are some valuable crisis radio frequencies regularly utilized all through the country:

NOAA Weather conditions Broadcast Frequencies:

• 162.4000 MHz

• 162.4250 MHz

• 162.4500 MHz

• 162.4750 MHz

• 162.5000 MHz

- 162.5250 MHz

- 162.550 MHz

FAMILY RADIO ASSISTANCE (FRS/GMRS) FREQUENCIES

The FRS/GMRS radio frequencies were embraced in 1996 to be utilized for family correspondence. Today, you know these frequencies as of now as "walkie-talkie" frequencies. You will not have the option to type in the specific recurrence, so utilize the all over bolts to test the recurrence until you see as the most grounded signal.

NOTE: The Baofeng UV-5R is significantly more impressive than a standard walkie-talkie, particularly on the off chance that you've redesigned it with another receiving wire or establishing. Imparting on these frequencies will probably over-power different transmissions. Remember this during a debacle situation.

WORLDWIDE PAIN RECURRENCE

The generally acknowledged, worldwide pain recurrence for any crisis radio transmission is VHF Channel 16 (156.800 MHz). Assuming you does not know what crisis recurrence to attempt and in the event that examining gives no transmissions entering into this recurrence is your most ideal choice. This recurrence is observed 24 hours daily by U.S. Coast

Watchman and oceanic staff all around the world. In the event that salvage tasks (land or ocean) are endeavoring to hail a crisis radio without really any information on the channel or arrangement, they will default to this recurrence.

TWO-METER BAND FREQUENCIES

Numerous neighborhood radio transmissions and repeaters work in the 2-meter band, or 144.000 MHz to 148.000 MHz. Check this scope of frequencies during a crisis, and you will probably contact others.

MULTI-UTILIZE RADIO/MURS CRISIS FREQUENCIES

MURS is an American VHF radio band, in no way related to FRS or GMRS. MURS basically fills the hole between the UHF frequencies given by FRS/GMRS, and the lower frequencies utilized by CB radios:

- 151.820 - Informal MURS calling recurrence

- 151.880 - Suggested repeater recurrence

- 151.940 - Crisis channel frequently utilized by preppers

- 154.570 - More seasoned business/business recurrence, still being used today

- 154.600 - More established business/business recurrence, still being used today

Other Valuable Crisis Radio Frequencies

- 156.050 - Port tasks

- 156.350 - Business use

- 156.450 - Boater calling

- 156.500 - General business

- 156.700 - Port tasks

- 156.850 - State and neighborhood government oceanic

- 157.000 - Port tasks

- 157.150 - U.S. Coast Watchman as it were

- 157.125 - U.S. Government as it were

- 161.825 - Public correspondence

INSTRUCTIONS TO PHYSICALLY PROGRAM A BAOFENG RADIO

Reasonable BaoFeng radios are inconceivably famous yet aren't easy to understand. It's a good idea for new proprietors to get a programming link and download the free Peep programming. It's a lot simpler than programming the hard way, and it offers extra capacities like naming diverts and switching off communicate for person on call frequencies you need to screen.

Yet, there are times when you might have to program a recurrence into your radio while in the field, so manual writing computer programs is an important expertise. Sadly, the included manual isn't exceptionally useful. There's an option accessible from the Chinese Radio Documentation Undertaking, yet it hasn't been refreshed in anywhere close to 10 years. Administrator KC7OM has made a printable reference card that you can print, cut out, and place between the battery and radio so it's consistently open. These directions are customized for the BaoFeng UV-5R, yet ought to work for a large portion of its subsidiaries like the BaoFeng BF-F8HP. Assuming that you're prepared to move up to a more straightforward radio to program, look at best handheld ham radios.

New to ham radio? Confounded by the phrasing? Look at our amateur's manual for ham radio.

Radio discussions occur on a recurrence, as 146.52. You're probably going to involve a few frequencies in radio correspondences, so rather than retaining an entire pack of frequencies, you can program them into your radio as channels. So in the event that you program 146.52 into channel 1, you simply need to make sure to tune into channel 1 rather than the recurrence. You can likewise look at your rundown of channels as opposed to entering them in without fail.

While conveying over a solitary recurrence that is called simplex. For instance, assuming you tune into 146.52 and another person is on that recurrence that you converse with, that is simple.

The method involved with adding frequencies to channels is called programming. This turns out to be more significant when you use repeaters, which are robotized stations that tune in for transmissions on one recurrence and "rehash" them over another recurrence. Repeaters have different intricacies also. Many require the radio to send an extraordinary tone, called a CTCSS tone or a DCS tone, before the repeater will rehash transmission from that radio. Except if you program the right tone into the channel, transmission to that repeater will be unbeneficial. Since repeaters utilize two frequencies for correspondences, that is called duplex, rather than simplex.

PROGRAMMING SIMPLEX CHANNELS INTO A BAOFENG

Physically programming simplex frequencies is less difficult than programming repeaters, so it's a decent spot to begin.

To start with, you should be in recurrence mode rather than channel mode. In recurrence mode, you dial in frequencies straightforwardly, while in channel mode you move between pre-customized channels.

Turn on your BaoFeng and check the screen out. Assuming that you see channel numbers at the right of the screen, you're in channel mode. To switch between the modes, press VFO/MR. In the event that you see a channel number on the right, you are in channel mode.

You've most likely seen two arrangements of frequencies on the screen, one on top and one on the base. The BaoFeng allows you rapidly to switch between two frequencies by squeezing the A/B button. To program a recurrence, you should be on the top recurrence, which is shown by a little bolt on the left. The bolt shows whether the upper or lower recurrence is chosen.

When you're in recurrence mode and you're on the top recurrence, enter the recurrence you need with the keypad, as 146.52. The BaoFeng takes three digits after the decimal point, so to enter a recurrence you want to add zeroes as far as possible, as so: 146.520. There's compelling reason need to enter the decimal point, so type 146520.

When the top recurrence is the one you need to program, press Menu. You can look to the MEM-CH menu thing however composing 27 is more straightforward. MEM-CH is the setting that projects channels into memory. When on MEM-CH, press Menu to change that setting. You realize you've done this effectively when the little bolt on the passed on moves from MEM-CH to CH-000, which is the default channel.

Tip: You can squeeze Exit whenever to leave the settings menu, either previously or in the wake of saving changes.

To enter a channel, either type it in on the console or utilize the bolt keys to look to it. When you have the channel you need to program the recurrence into, press Menu followed by Exit.

When back on the principal menu, press VFO/MR to change to channel mode. Either look to the modified channel with the bolts or enter it on the keypad to affirm that the channel was customized effectively.

PROGRAMMING REPEATERS INTO A BAOFENG

When you have the rudiments of programming channels, you're prepared to add repeaters physically. For repeaters, you want four snippets of data: the primary recurrence, the offset, offset heading, and the tone. (On RepeaterBook, tone is recorded as tone in/tones out, yet they ought to be no different for FM stations viable with the BaoFeng.) The principal recurrence is the one the repeater communicates on, which is the recurrence you pay attention to. The offset is the recurrence that the repeater tunes in on, which is what you communicate to. Balances are communicated in sure or negative numbers. So on the off chance that the repeater recurrence is 146.67 MHz and the offset is 0.6 MHz, you communicate on 146.07. Assuming the offset was +0.6, you would rather send on 147.27. For instance, we should utilize W4CAT, which is a repeater around Nashville:

• Recurrence: 146.955

• Balance: - 0.6 MHz

• CTCSS: 114.8

Rehash the primary two or three stages from a higher place: ensure you're on the top recurrence in VFO mode and enter the repeater recurrence.

Set the offset heading:

1. Press Menu

2. Type 25 or look to SFT-D

3. Press Menu

4. Use the bolts to set +, - , or off

5. Press Menu

Then set the offset recurrence:

1. If you're in the menu, press 26 or look to Counterbalance

2. Press Menu

3. Enter the offset (for 0.6, type 000600)

4. Press Menu

At last, the tone:

1. If you're in the menu, press 13 or look to T-CTCS (short for send CTCSS)

2. Press Menu

3. Use the keypad to enter the tone recurrence

4. Press Menu

SETTING THE CTCSS TONE

Whenever everything is set accurately, save the recurrence to a channel similarly as you would for simplex. The offset course, offset recurrence, and CTCSS tone settings ought to save to that channel. It's not difficult to check whether everything was saved accurately. Look to the repeater's channel. It would be ideal for you to see +-show up at the highest point of the screen to demonstrate an offset. At the point when you press the PTT button as an afterthought to send, you ought to see two things: CT shows up on the left, demonstrating a CTCSS tone being sent, and the recurrence dropping or expanding to the offset. For example, for 146.955 and an offset of - 0.6 MHz, the recurrence changes to 146.355.

At the point when an offset is set, the recurrence naturally changes when you send.

To set a DCS tone, you follow a similar strategy, with the exception of you set menu thing 12, T-DCS, all things being equal.

STEP BY STEP INSTRUCTIONS TO ERASE A CHANNEL

Erasing a channel than add one is a lot more straightforward:

1. Press Menu

2. Enter 28 or look to DEL-CH

3. Press Menu

4. Scroll to or enter the channel to erase

5. Press Menu

Be cautious here, since there is no affirmation brief. In the event that you get into the menu and adjust your perspective, press Exit before stage 5.

THE END